I0470662

National Guard Plane Crash at Hotel Site
Evansville, Indiana

Investigated by: Mark Chubb

This is Report 064 of the Major Fires Investigation Project conducted by TriData Corporation under contract EMW-90-C-3338 to the United States Fire Administration, Federal Emergency Management Agency.

FEMA

Department of Homeland Security
United States Fire Administration
National Fire Data Center

U.S. Fire Administration Fire Investigations Program

The U.S. Fire Administration develops reports on selected major fires throughout the country. The fires usually involve multiple deaths or a large loss of property. But the primary criterion for deciding to do a report is whether it will result in significant "lessons learned." In some cases these lessons bring to light new knowledge about fire--the effect of building construction or contents, human behavior in fire, etc. In other cases, the lessons are not new but are serious enough to highlight once again, with yet another fire tragedy report. In some cases, special reports are developed to discuss events, drills, or new technologies which are of interest to the fire service.

The reports are sent to fire magazines and are distributed at National and Regional fire meetings. The International Association of Fire Chiefs assists the USFA in disseminating the findings throughout the fire service. On a continuing basis the reports are available on request from the USFA; announcements of their availability are published widely in fire journals and newsletters.

This body of work provides detailed information on the nature of the fire problem for policymakers who must decide on allocations of resources between fire and other pressing problems, and within the fire service to improve codes and code enforcement, training, public fire education, building technology, and other related areas.

The Fire Administration, which has no regulatory authority, sends an experienced fire investigator into a community after a major incident only after having conferred with the local fire authorities to insure that the assistance and presence of the USFA would be supportive and would in no way interfere with any review of the incident they are themselves conducting. The intent is not to arrive during the event or even immediately after, but rather after the dust settles, so that a complete and objective review of all the important aspects of the incident can be made. Local authorities review the USFA's report while it is in draft. The USFA investigator or team is available to local authorities should they wish to request technical assistance for their own investigation.

This report and its recommendations were developed by USFA staff and by TriData Corporation, Arlington, Virginia, its staff and consultants, who are under contract to assist the USFA in carrying out the Fire Reports Program.

The USFA greatly appreciates the cooperation received from Chief Douglas Wilcox and members of the Evansville Fire Department and from Director Sherman Greer, Evansville-Vanderburgh County Emergency Management Agency

For additional copies of this report write to the U.S. Fire Administration, 16825 South Seton Avenue, Emmitsburg, Maryland 21727. The report is available on the USFA Web site at http://www.usfa.dhs.gov/

U.S. Fire Administration

Mission Statement

As an entity of the Department of Homeland Security, the mission of the USFA is to reduce life and economic losses due to fire and related emergencies, through leadership, advocacy, coordination, and support. We serve the Nation independently, in coordination with other Federal agencies, and in partnership with fire protection and emergency service communities. With a commitment to excellence, we provide public education, training, technology, and data initiatives.

 FEMA

TABLE OF CONTENTS

National Guard Plane Crash At Hotel Site
Evansville, Indiana
February 1992

Local Contacts: Chief Douglas Wilcox
Deputy Chief Thomas Gronotte
Investigator Jesse Storey
Investigator Roger Griffen
Evansville Fire Department
Civic Center Complex, Room 203
Evansville, Indiana 47708
(812) 426-5668

Sherman Greer, Director
Evansville-Vanderburgh County
Emergency Management Agency
City-County Complex, Room 18
Evansville, Indiana 47708
(812) 426-5603

OVERVIEW

At 9:53 a.m., February 6, 1992, a Kentucky Air National Guard C-130B military transport plane crashed at the site of a hotel and restaurant complex while performing routine pilot proficiency exercises at Evansville Dress Regional Airport. The crash and resulting fire killed all five crew members and 11 civilians on the ground – nine in the hotel and two in the adjacent restaurant. In addition to those who died at the scene, one of the emergency responders, an Evansville Police Officer, died February 24, as a result of injuries he sustained working at the crash site. Emergency responders from the City of Evansville, the Evansville Regional Airport, and police, fire, and emergency medical service (EMS) agencies across Vanderburgh County responded to the disaster.

SUMMARY OF KEY ISSUES

Issues	Comments
Casualties	17 dead, including five aircraft crew, nine in hotel, and two in restaurant. A police officer died 16 days later from injuries incurred during rescue operations. 15 civilian injuries and 6 emergency responder injuries were also reported.
Building Construction	Well designed, constructed, and maintained building features limited fire spread and loss of life.
Incident Command System (ICS)	Use of the ICS throughout the operation contributed to effective command, control, and communication among diverse emergency and disaster response agencies and a smooth transition from response to recovery.
Emergency Planning and Preparedness	Routine emergency planning and preparedness activities such as planning and exercises, augmented by regular emergency responder training facilitated coordination of structural firefighting and aircraft rescue and firefighting (ARFF) forces, EMS, law enforcement, public works, coroner's office, and military authorities.
Occupant Response	Many occupants took self-protective actions. Three unrelated individuals sought refuge together in a room rather than evacuating the building. Some occupants reportedly exited, then reentered to rescue occupants trapped on upper floors.
Media and Public Relations	The dissemination of information through the media was timely and effective. Emergency management officials took advantage of media interest in the crash to direct family and concerned citizens away from the scene to a citizens' center where they could obtain information about those injured and killed.
Critical Incident Stress Management	Civilian and military critical incident stress debriefing teams and social service agencies responded to manage the emotional and psychological impacts of the incident on responders, citizens, and family members.

A similar incident involving an Indiana Air National Guard A-7D attack aircraft occurred in October 1987 in Indianapolis. In that incident, the aircraft struck a Ramada Inn hotel killing nine employees who were inside the building.[1] The pilot safely ejected prior to impact. Like the Indianapolis incident, the crash in Evansville occurred in mid-morning during a routine training mission. Both incidents also involved Air National Guard aircraft. However, the similarities do not end there.

In both cases, well designed, constructed, and maintained buildings minimized fire spread, structural damage, and most likely the loss of human life. Clearly, investments in built-in fire protection pay off, even in such unforeseen events as aircraft crashes outside the building, or literally into the side of the building, as was the case in Indianapolis.

In addition to the building performance, the effectiveness of emergency responders in controlling these incidents and coordinating massive response efforts demonstrated the value of training, planning, and preparedness. The use of ICS in each case significantly aided coordination, and in Evansville helped provide for a smooth transfer of authority to disaster management and later to military officials, which clearly expedited investigation and recovery efforts.

---

[1] *Ramada Inn Air Crash and Fire, Wayne Township, Indiana* (October 20, 1987), Report 014, Major Fires Investigation Project, Technical Report Series, Emmitsburg, MD: Federal Emergency Management Agency, United States Fire Administration.

Table 1. Similarities in Evansville and Indianapolis Airplane/Hotel Crashes

- Occurred mid-morning with relatively few occupants in hotel

- Well constructed and maintained buildings minimized fire spread and structural damage

- Rapid fire and emergency service response

- Effective implementation and use of disaster plan

- Effective use of ICS

- Effective integration and use of mutual aid resources

- Effective media relations and use of media to disseminate information about the incident

Communication among emergency responders is viewed as critical at any emergency scene, but in both of these cases communication with the public was viewed as equally important; public information activities were an integral part of the disaster response. In Evansville, media briefings were used to direct family members and concerned citizens away from the accident site to a community center staffed by local clergy and counselors, where they could obtain information about deaths and injuries.

The Evansville incident provides an opportunity to study the behavior of building occupants in response to a highly unusual and virtually unforeseeable event and to learn how individuals coped with the unusual and confusing circumstances which confronted them. The factors of human response to fire situations are often overlooked in the development of codes, standards, and regulations that are intended to provide public safety.

BUILDINGS AND CRASH SITE

The Drury Inn hotel and Jojo's Restaurant were located in separate buildings which occupied a common parcel of land at the intersection of Lynch Road and U.S. 41 less than one mile south of Evansville Dress Regional Airport. A large undeveloped agricultural tract was situated east and south of the crash site. Diagrams of the buildings and site appear in Appendix A.

The four-story guest room portion of the hotel was completed in July 1980. The building was of noncombustible, Type 2 one-hour construction as classified by the *Uniform Building Code*, 1988 edition, upon which the Indiana Building Code is based. The restaurant was of unprotected Type 2 construction. In 1989, the motel was renovated and a four-story atrium lobby was added at the west end, separated from the guest wing by a two-hour fire wall with automatic closing fire doors.

Guest rooms were separated from the corridors and from one another by one-hour fire-resistance rated walls. Guest room doors were self closing twenty-minute fire rated doors in fire rated door frames. Where doors remained closed during the incident, they prevented the fire from spreading into the corridor.

The motel was protected throughout by an automatic fire alarm system. Smoke detectors and manual stations in the corridors initiated horn/strobe evacuation signals also located in the corridors. The activation of the fire alarm system caused the automatic closers to release the doors protecting the atrium firewall openings at the west end of the corridors.

Mechanical air handling equipment was designed to shut down upon detection of smoke in the supply or return air. Consequently, recirculation of smoke did not appear to be a significant problem during this fire.

Single station 110-volt AC-powered smoke detectors were installed in each guest room. They had little impact on the outcome of this incident since the first detectors activated were in the corridors, probably on the fourth floor.

A standpipe system was provided for both occupant and fire department use. Risers with fire department hose outlets were located in each of the three exit stairways. Occupant-use hose stations were provided outside the stairway enclosures. Firefighters used standpipes during the incident to battle the fire inside the hotel. However, initial attack lines were drawn directly from the pumpers.

Automatic sprinklers connected to the domestic water system protected designated hazardous areas, i.e., workshop, laundry room, and storage rooms, on the first floor. The atrium was also protected by a partial automatic sprinkler system. None of the sprinklers activated during this incident.

Emergency lighting and internally illuminated exit signs and were installed in the corridors to assist occupants in locating exits. However, dense smoke from the burning aviation fuel probably obscured the signs and impeded the usefulness of the emergency lighting.

THE CRASH AND FIRE

On the morning of February 6, 1992, a five member crew from the Kentucky Air National Guard's 123rd Tactical Airlift Wing based at Standiford Field in Louisville, Kentucky, was performing touch-and-go landings at Evansville Dress Regional Airport as part of routine pilot proficiency training. The crew consisted of an experienced instructor pilot, two co-pilots, a flight engineer, and a loadmaster.

The type of military airplane – the Lockheed C-130-B *Hercules* – which crashed in Evansville on February 6, 1992, was renowned for its safety record and reliability. C-130 crashes are rare events. Historically, most aircraft crashes occur during takeoffs and landings.

At the time of the crash, one of the co-pilots was flying the aircraft under the supervision of the instructor pilot. According to U.S. Air Force investigators, the crash was attributed to pilot error, which produced an unrecoverable stall following a low level approach maneuver over the airfield. A stall results from insufficient airflow under the wing of an aircraft. The stall was the result of insufficient airspeed at the time a turning maneuver was executed. Without sufficient airspeed or altitude (approximately 1,300 feet), the crew was unable to regain control of the aircraft before it crashed.

The aircraft went down approximately one mile south of the departure end of Runway 22 at 9:53 a.m., impacting in the courtyard behind the Drury Inn and Jojo's Restaurant. The aircraft altitude and rate of descent produced a very small crash impact zone. At the time of impact, the aircraft was descending almost vertically at a rate of between 4,500 and 6,000 feet per minute with the nose elevated 4-degrees above vertical and the right wing 47-degrees below horizontal. The impact created a crater 8-feet deep and 12-feet across. The immense force of the impact splashed burning aviation fuel toward the hotel and broke windows across the center portion of that building. A large piece of the tail section of the aircraft landed on the rear quarter of the restaurant causing it to collapse, pinning two victims. Part of the skin of the right wing was propelled over the hotel into the parking lot south of the hotel and large chunks of the concrete pool deck and airplane parts landed on the four-story roof of the hotel.

The fireball created by the crash impinged directly against the center portion of the north wall of the Drury Inn. Windows broken by the force of the impact and the radiant heat allowed the fireball to spread into several hotel rooms on all four floors. However, fire spread beyond these rooms was minimal in most instances because of fire-resistive construction separating the guest rooms from the corridors. In the few locations where the fire did extend beyond a guest room, doors had been left open by guests or hotel housekeepers. Fire extension in the corridors was minimal due to the limited fuel loading and interior finishes.

The fire was accompanied by choking black smoke produced by the burning aviation fuel. The high concentration of aviation fuel present in the air after the airplane broke apart made the smoke particularly dense and acrid due to incomplete combustion of the fuel.

The hotel operators believe that seventy-five to eighty people were inside the building at the time of the fire. This number included 11 employees of a local plumbing supply company and two instructors from the University of Southern Indiana who were conducting a total quality management seminar in a fourth floor meeting room. The post-crash fire killed nine of the people in this meeting room and seriously injured the other four. In addition, 11 other hotel occupants sustained various injuries, mostly from smoke or toxic fume inhalation.

At the Jojo's Restaurant, two employees were killed when part of the airplane impacted directly on the kitchen area in the southeast quarter of the building. Two workers, a waitress, and a dishwasher, were pinned in the wreckage of the collapsed portion of the structure. Approximately twenty-five people escaped from the restaurant unharmed or with minor injuries.

EMERGENCY RESPONSE

Firefighters at Evansville Fire Station #2 heard the explosion from the crash and went outside to investigate. They could see the smoke plume rising from the crash site, and, at 9:54 a.m., Engine Company 2 reported that they were responding to an explosion and fire which they believed was in the vicinity of the rail yard west of U.S. 41. (A few weeks prior to the crash, Evansville firefighters responded to a large hazardous materials incident at the rail yard and may have believed this type of incident was recurring.)

At the same time, air traffic controllers at the Evansville Tower alerted Airport Safety Department crews, who responded with one crash truck, Rescue 1.

Firefighters from both departments met Vanderburgh County Sheriff's deputies on the scene. A deputy had been refueling his cruiser at the service station across the street from the hotel and witnessed the crash. Other deputies responded from a Sheriff's substation office approximately one-quarter mile north of the crash site.

Other emergency responders, including several Evansville police officers and EMS units from the local contract ambulance services also responded on the initial report of an aircraft down. (See Appendix B for a list of all the agencies which responded to this incident.)

The first firefighting units arrived on the scene less than two minutes after the airplane crashed. Evansville Engine Company 2 positioned itself at the east side of the crash site and advanced hoselines to protect the exposure along the north wall of the hotel, before initiating rescue efforts inside the hotel. Police officers, sheriff's deputies, other emergency responders, and some hotel and restaurant occupants also initiated independent efforts to reach people trapped inside the burning hotel and restaurant. Meanwhile, Rescue 1 began attacking the fuel fire from the northwest side using its on-board foam/water supply.

The dense smoke made approaching the airplane and entering smoke-filled corridors particularly hazardous to personnel not wearing self-contained breathing apparatus (SCBA). An Evansville police officer, a Vanderburgh County sheriff's deputy, and an emergency medical technician (EMT) who were not wearing SCBAs sustained serious respiratory injuries during this incident. The Evansville police officer entered the building several times but did not complain of any injuries until several hours later. He was admitted to an area hospital later in the day and died on February 24, due to complications from his injuries. The other two responders were exposed outside the building and were less seriously injured but did require treatment at the hospital.

The Evansville Fire Department initially dispatched four engine companies (including Company 2), one aerial ladder, one heavy rescue company, and one district chief. At 9:57 a.m., while still en route to the scene, the Evansville District Chief requested a second alarm. The second alarm brought three additional engine companies, one aerial ladder truck, one rescue company with hazardous materials equipment, and one battalion chief. Additional administrative personnel responded from Fire Headquarters upon the report of an airplane down. A third aerial ladder and a quintuple combination engine company were special called to bring the total number of units at the scene to seven engine companies, four ladder companies (including the quint), one ARFF vehicle, two rescue companies, and two district chiefs. A number of other firefighters and township fire units also responded, although they had not been requested by Evansville authorities.

Upon his arrival, the first district chief on the scene established a Command Post and activated the ICS. The Command Post was set up in front of the hotel on U.S. 41 west of the crash site. An interior sector was established to direct rescue efforts inside the hotel, and another sector was established to oversee operations at the east end of the crash site. The responding companies were committed primarily to search and rescue operations. Most of the exterior fire was knocked down by the initial attack and the interior fire was controlled with handlines.

Approximately 15 to 20 minutes into rescue operations, while the fire on the north side of the building was being controlled by the airport ARFF crew using foam and by structural firefighters using water hoselines, an aerial ladder was used to extricate three people who had taken refuge from the fire in room 405, on the south side of the fourth floor. Two of those rescued were severely burned and were later identified as having been in room 416 where most of the fatalities occurred when the airplane impacted. The other person, who was not injured, was the guest registered to room 405.

The fire was declared under control at 11:44 a.m., less than two hours after the initial crash.

Within 10 minutes of the crash, at 10:03 a.m., the Evansville Vanderburgh County Emergency Management Agency activated its Emergency Operations Center (EOC). Once the center was operational, the agency director responded to the crash site to support fire control and rescue operations and to assume command of recovery operations once the fire was controlled. Overall command of the incident was transferred to the Evansville-Vanderburgh County Emergency Management Agency at approximately 12:15 p.m.

EMS RESPONSE

A total of 34 ambulances from 12 service providers were sent to the scene to treat and transport victims of the crash. The local ambulance service which provides EMS for the City of Evansville, activated its Disaster Alert Plan which placed the three area hospitals on alert and summoned assistance from all available EMS from the six township volunteer departments in Vanderburgh County. The EMTs were instructed to respond to an EMS staging area near the crash site. A total of 83 EMS personnel responded to the incident.

COMMUNICATIONS

The Evansville Police and Fire Departments had completed conversion to a new communications system shortly before the incident. The new system consisted of enhanced 9-1-1, computer-aided dispatching, and an 800-MHz trunked radio system. The crash served as the first major test of the new communications equipment and procedures. According to local officials, both passed with flying colors.

Despite the large number of calls generated in response to the crash, the new system permitted dispatchers to keep pace with events without losing track of field operations. During the first hour of the incident, the dispatchers handled more than double the normal call volume. The second hour produced an even more dramatic 500 percent increase as calls poured in from outside the local area, many from the Regional and National media. Call volume remained high for more than 10 hours following the crash.

Figure 1.Total Calls Processed by Evansville Central Dispatch
February 6, 1992 (6:00 a.m. to 8:00 p.m.)

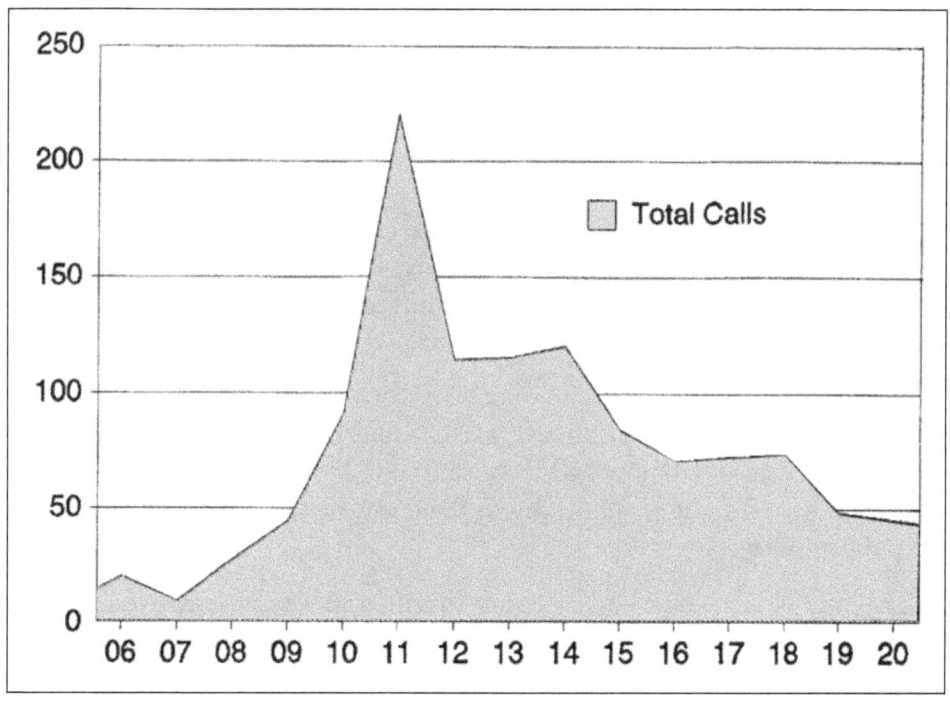

Source: Evansville R-e Department

While the dispatchers were managing the high volume of incoming calls and notifications associated with the incident, radio communications at the scene were aided by the trunked 800-MHz radios, which work much like cellular telephone service. This type of radio system permits a large number of users to share a relatively small number of common frequencies. Users are divided into talk groups which share similar responsibilities or communications needs. Table 2 lists the talk groups used by Evansville and Vanderburgh County agencies during this incident. The shaded boxes indicate talk groups which were used for interagency communications.

Other agencies involved in rescue operations, including the EMS providers and airport ARFF crews were not equipped with 800-MHz trunked radios and encountered difficulty communicating effectively with the Incident Commander. These agencies operated on their own frequencies during the incident and had to relay their messages through liaison officers detailed to the command post or through face-to-face communication with Evansville or Vanderburgh county officials.

Table 2. City of Evansville Central Dispatch Talk Groups

Evansville Fire Department	Evansville Police Department	Vanderburgh County Sheriff's Office
Fire Dispatch	Police Dispatch	Sheriff's Office Dispatch
Fireground 1	Tactical 1	
Fireground 2	Tactical 2	
Administration	Criminal Investigation Division	Information
Investigators/Inspectors	Information	
Disaster 1		
Volunteer Fire Dispatch	Disaster 2	
C^3 (Command)		
Telephone Patch		

MEDIA RELATIONS

Within 20 minutes of the crash, calls began coming into the EOC from National radio and television networks, large-market metropolitan television stations, wire services, and newspapers requesting information about the incident. At the scene, television camera crews from local television stations, newspaper reporters and photographers, and local wire service correspondents were vying to capture their respective angles on the story.

Perimeter security at the scene minimized media interference with emergency operations. Nonetheless, the media's need-to-know had to be satisfied. Moreover, it became apparent, with the onslaught of bystanders and sightseers flocking to the crash site, that the media could be helpful in informing the public not to come to the scene.

At 11:45 a.m., the fire chief, police chief, county sheriff, and a representative of the Indiana State Police held an initial news briefing at the incident scene. Each official provided a brief overview of the current status of their response, including the initial casualty figures, progress of fire suppression efforts, and information about traffic control around the incident scene.

A formal press briefing center was set up in the Ramada Inn located across the street from the incident site at about 12:15 p.m. A second press briefing was held at 1:00 p.m. to update the media and attempt to control rumors about the number of dead and injured. Representatives from the Evansville Police and Fire Departments, Indiana State Police, Vanderburgh County Sheriff's Office, county coroner's office, State Emergency Management Agency, Alexander Ambulance Service (the local EMS contractor), and the local Emergency Management Agency were present at the news conference. Later news conferences were attended by representatives of the Kentucky Air National Guard and U.S. Air Force Safety Investigation Board investigating the crash.

A total of eight press briefings were held in the three days following the crash, including five on the first day. During the briefings on the afternoon of the day of the crash, information was provided through the media to direct people to a community center set up to handle public inquiries about the status of family members and to provide relief assistance and counseling to families of crash victims.

CRITICAL INCIDENT STRESS DEBRIEFING

Both local authorities and the military recognized the extreme emotional and psychological impact the crash could have on emergency responders and the community at-large. Critical incident stress debriefing (CISD) specialists from the Southwest Indiana Critical Incident Stress Management Team and Scott Air Force Base (AFB) Medical Center responded to the scene to assist those affected by the incident. Debriefing sessions were held for both the emergency responders and the community. CISD team members reported that these sessions were well attended.

The CISD intervention was intended to prepare those who responded to or witnessed the incident for the possible physical, emotional, and cognitive effects to the stressful events they had experienced. The sessions provided useful information to the participants on how they themselves could minimize or manage their stress responses. They also identified agencies and individuals in the community who could provide expert assistance to those experiencing severe reactions to the incident. (See Appendix C for more information on the content of the critical incident stress debriefing program.)

According to the specialists sent from Scott AFB, this was the first time military critical incident stress teams have responded to a non-combat incident.

POST-INCIDENT CRITIQUE

On February 19, 1992, the Evansville-Vanderburgh County Emergency Management Agency conducted an after-action critique of response and recovery operations. All of the agencies involved were invited to send representatives. (See Appendix D for a copy of the after-action-assessment comments of the critique participants.) Overall, the feedback from all agencies was positive. However, several areas for improvement were noted. These included:

- Better communication is needed between the incident scene and the hospitals regarding casualties;

- Safety and accountability of emergency responders were compromised by freelancing by individuals and by the uncoordinated actions of self-dispatched agencies and personnel and by the lack of a unified accountability system;

- Scene security was established early but compromised by agencies and responders who had not been requested by local officials and were not integrated into the ICS;

- Better visual identification is needed for Command Post and Incident Commander for reference by agencies not on the local radio net;

- Radio communications with outside agencies, including EMS responders and Indiana State Police, were compromised by lack of common radio net or channel; and,

- Better coordination with the news media was needed in light of the overload of information requests and some misinformation provided in early reports.

Most agencies agreed that, notwithstanding a few minor problems, communication and cooperation among the diverse groups involved in the incident were exceptionally good. Moreover, they commended the local Emergency Management Agency on its excellent response during its first major disaster.

OCCUPANT RESPONSE

Human behavior in fire emergencies can have a significant impact on what firefighters encounter when they arrive on the scene. Firefighters rarely have an opportunity to find out what the occupants of a building were doing before they arrived. This incident presented a unique opportunity to evaluate what many of the occupants did before the arrival of the fire department and how it impacted the loss of life and the rescue situation confronting fire service and other emergency responders at this incident. The information which follows is based on press accounts, confirmed through interviews with fire officials and others who investigated this incident and interviewed those present when the airplane crashed.

Room 416

Thirteen employees of a local plumbing supply company were attending a seminar in a fourth floor meeting room (416) when the aircraft crashed. (A diagram of the room is shown in Figure 2.) Four of these people died instantly when the windows blew out and an expanding ball of burning aviation fuel engulfed the room. Five others lived long enough to take other actions. Three of them, their access to the exit blocked by the fire, sought refuge in the bathroom. A fourth went to the telephone and tried to call for help. The fifth tried to reach the door to the corridor but was felled by the smoke and heat before he could reach it. These nine victims were the only people to die inside the hotel. Four other occupants were seriously injured but managed to escape the inferno in room 416. What happened to these four individuals affected what the firefighters saw when they arrived.

Figure 2. Detail of Room 416

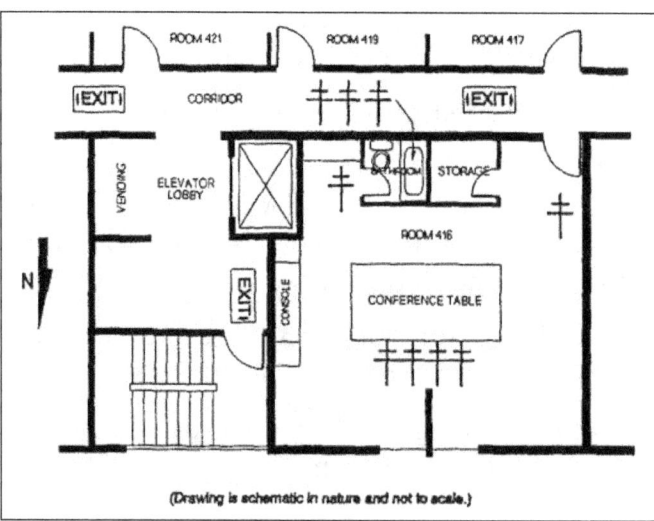

(Drawing is schematic in nature and not to scale.)

Immediately prior to the crash, the two instructors and 11 seminar attendees in room 416 had just reconvened after a short break. One of the instructors was writing on an easel pad at the front of the room. The other instructor was standing near her at the front (west end) of the room, closer to the door. The seminar participants were seated around a conference table in the center of the room. As the presentation was about to resume, the aircraft impacted directly below the windows of room 416.

After the fire, the other instructor, recalled, "I looked at Lynn and behind her out the small window. I didn't see the plane crash. All I saw was this ball of fire which almost looked like it was coming at us in slow motion."[2] In another interview, he recounted, "There was a strange airplane noise. Then a crash…there was a ball of fire coming at us."[3]

The four occupants (including the two instructors) who were nearest the doorway escaped into the corridor.

The first occupant believed to have escaped room 416, the instructor who witnesses the fire through the window, later recounted, "I just thank God for my Cub Scout training. I knew I must be on fire, so I just dropped to the floor and rolled."[4] After extinguishing the fire on his clothing, the instructor ran west down the corridor, past the turn toward the west stairway, and onto the balcony overlooking the lobby atrium. There he yelled down to occupants and staff below that there was a fire in room 416, and that twelve people were inside the room. (He was apparently unaware that others had escaped.)

Moments later, the fire alarm system activated, releasing the magnetic door holders keeping the balcony door open. When the door closed, he was trapped on the small balcony, four stories above the atrium. The fire door protected him from the fire's effects but prevented his reentry to the corridor because there was no hardware on the atrium side. Firefighters judged that he was not in immediate danger and attended to other occupants in more danger before rescuing him about one hour after the crash.

The instructor writing on the easel pad and one of the seminar participants also proceeded west down the corridor. As they approached room 405, prior to reaching the stairway, its occupant opened his door to investigate. The man in room 405 quickly realized that the building was on fire and that the corridor had become unsafe and recognized that both the instructor and seminar attendee had been injured and were in distress. The three retreated into room 405, went to the window on the south side to summon help, and while waiting to be rescued, sealed the undercut of the doorway with towels to minimize smoke infiltration. All three were rescued by Evansville firefighters using an aerial ladder 15 to 20 minutes after the crash.

The actions of the fourth occupant to escape room 416 are less clear. The occupant of room 419 (directly across the hallway from room 416), reported that when she entered the hallway and began crawling toward a light at the end of the hallway (probably at the east end), she encountered another woman who was burned and in severe pain. She later stated, "I told her (the woman in the hallway) to get down and crawl. She said we'd never get out and I said, 'Yes, we will; I'm not gonna die in here.' Just follow the light."[5]

---

[2] "Drury Inn burn victim dodged 'ball of fire' to survive, "The Evansville Press, Feb. 8, 1992, p. 2

[3] "Survivors mourn, relive nightmare," Evansville Courier, Feb. 8, 1992, p. A1.

[4] "Search for crash cause begins," The Evansville Press, Feb. 7, 1992, p. 1.

[5] "Survivors mourn, relive nightmare," Evansville Courier, Feb. 8, 1992, p. A1.

Firefighters found three victims in the bathroom in room 416 huddled under a running shower. The occupants had turned on both the hot and cold water in an effort to protect themselves, but all had succumbed to smoke inhalation by the time rescuers reached them.

The woman from room 419 reported hearing their plaintive cries for help as she herself fled to safety.[6] A fourth victim was found slumped on the floor just outside the bathroom where it appeared he had tried to use the telephone to call for help. One other occupant appeared to have tried to escape the room but collapsed and died a few feet from the doorway to the corridor.

Drury Inn Lobby[7]

Two Drury Inn guests reported that they were eating breakfast in the hotel lobby when the plane crashed. Both reported hearing aircraft noise prior to impact. (They were probably in the lobby when the first instructor to escape room 416 yelled down from the balcony that the building was on fire.) They exited the building through the south exit doors to the parking lot. Once outside, they heard calls for help from occupants on the fourth floor (possibly the occupants of room 405). They were joined in the parking lot by a bystander, who had run to the scene from a nearby business. The three men then entered the building and rescued several occupants in two successive attempts before worsening heat and smoke conditions prevented them from continuing. They suspended their rescue efforts when firefighters arrived on the south side of the building.

JoJo's Restaurant[8]

Two men were working in the kitchen of JoJo's Restaurant when they felt the impact of the crash. One of them reported the fire cut off access to the nearest exit. The other was standing in front of the grill and fryers when they lunged forward as the ceiling caved in from the crash impact. Both men assisted in efforts to reach two of their co-workers, a waitress and a kitchen worker, who were pinned in the structure's wreckage. The second man estimated in an interview with reporters that they worked unsuccessfully for about 15 minutes to rescue the others. The two people pinned in the wreckage died.

Evansville fire officials estimate that as many as 25 employees and customers escaped unharmed from JoJo's Restaurant, using the front entrance which was on the opposite side of the building from the fire.

ANALYSIS

This fire presents an opportunity to learn lessons about several significant issues, including building construction and exit design, emergency response, incident command, communications, and human behavior in fire emergencies.

---

[6] "Survivors mourn, relive nightmare," Evansville Courier, Feb. 7, 1992, p. A1.

[7] "Business partners staying at Drury recall crash," The Evansville Press, Feb. 7, 1992, p. 3.

[8] "Restaurant workers first to feel touch of tragedy," The Evansville Press, Feb. 6, 1992, pp. 1, 14.

Building Features

The noncombustible construction and relatively low fuel load from interior finishes and furnishings kept fire damage to a minimum. Most of the lethal fire and smoke production appeared to have resulted from burning aviation fuel. Only north facing rooms, which were directly exposed to the fire outside, sustained thermal fire damage. Smoke spread was heavy only where doors were left open by occupants, such as the fourth floor where the doors to rooms 416 and 422 were left open. Very little smoke or fire damage was noted in corridors on the first through third floors and none of the rooms on the south side of the building sustained thermal fire damage. The automatic-closing fire doors at the west end of the corridors worked in concert with the fire alarm system to prevent the fire from spreading into the atrium, especially on the fourth floor where open room doors permitted significant volumes of fire and smoke to enter the corridor.

Notwithstanding these successes, this fire points out other areas where building features, although code-conforming, may have contributed to injuries and loss of life. For example, none of the Nationally recognized model codes would have required two exits from room 416. Nonetheless, two remote means of egress from room 416 probably could have prevented the loss of at least four lives.

Emergency Response

Unrequested mutual aid response from neighboring departments and off-duty firefighters complicated incident command, and compromised fireground accountability and security of the incident scene. With law enforcement personnel assisting firefighters and attempting several risky rescues without proper protective equipment, early fire scene security was not established.

Fire department personnel cordoned-off the perimeter of the crash site themselves approximately 30 minutes after the crash. As sufficient firefighting and rescue personnel arrived on the scene, law enforcement efforts were gradually redirected to monitoring scene access and controlling traffic.

Without early control of the fire scene, effective personnel accountability was nearly impossible. One Evansville police officer who had engaged in rescue efforts prior to the arrival of additional firefighters died on February 24 of injuries he sustained in the line-of-duty. A sheriff's deputy and an EMT were also seriously injured but later recovered from smoke inhalation and exposure to toxic products of combustion sustained because they were not using appropriate respiratory protection. Four other emergency responders sustained minor injuries, including sprains, strains, and bruises.

Emergency Management

The participants in this incident generally agreed that the use of the ICS served them well in coordinating the emergency response and ensuring a smooth transition to recovery operations.

The timely activation of the joint city/county EOC aided in the smooth and effective management of this incident. Incoming requests for information about the incident were effectively managed while outgoing notifications and arrangements for assistance were executed. This extended as far as coordinating the response of clergy and counselors to assist victims' families and emergency responders, which helped expedite the recovery process.

By responding to the incident site, the Emergency Management Director facilitated a smooth transition from emergency response to the investigation and recovery phases.

By establishing good media relations, the police, fire, and emergency management personnel were able to successfully manage the intense public interest in this incident and prevent the public from unwittingly exposing themselves to harm and creating a larger problem for rescuers.

Communications

Evansville's new 800-MHz trunked radio communication system provided needed flexibility and enhanced communications effectiveness during this very complex incident. Although call volume and radio traffic far exceeded reasonably anticipated emergency demand levels, communications personnel managed the situation admirably and gave much of the credit to the new system and procedures.

Personnel at the fire scene were impressed by the extent to which the trunked radio system allowed them to maintain good interagency communication. The only negative comments about fire scene communication among responders only served to highlight the value of the new equipment: the agencies that were not part of the new system now want to be added to the network.

Occupant Response

By the time firefighters arrived on the scene, the occupants of the Drury Inn and Jojo's Restaurant who survived the plane's initial impact had already fled the fire, were seeking refuge from it, or were attempting to save others who were trapped. Firefighters and other emergency responders encountered many of these people during their initial operations. The three people who sought refuge together in room 405 were extricated using an aerial ladder 15 to 20 minutes after the crash. The instructor who ran to the fourth floor balcony and became trapped there was rescued about an hour after the crash. Other rescuers tried in vain to reach the waitress and kitchen worker pinned in the structural debris of the Jojo's Restaurant.

Most of the survivors of the Drury Inn and Jojo's Restaurant evacuated without assistance. Those who were unable to escape were generally overcome by the fire, pinned in debris, or cut off from their means of escape. Although eleven people were killed in the hotel and restaurant, many more lives could have been lost were if not for the prompt, orderly evacuation of most of the occupants. Those who did not evacuate stayed to help others who they felt were in greater need of assistance than themselves, such as the two lobby occupants and the kitchen workers at Jojo's. Their actions demonstrate that people are far more likely to respond adaptively and altruistically in the face of danger than to panic.

Those who attempted to flee but did not successfully evacuate the building illustrate a phenomenon which is common in multiple fatality fires. Three occupants sought refuge in the bathroom in room 416 because they could not reach an exit. Two others sought refuge with a third occupant and protected that room until rescued. In one case, these actions were successful; in the other they were not. In both cases, the occupants were attempting to protect themselves by placing distance or a barrier or both between them and the fire. In the absence of a better course of action or a clear objective to evacuate the building, people will employ alternatives which are less than optimal. This underscores the need for clear, simple, direct means of egress and routine evacuation planning.

Means of Egress Signs and Lighting

Emergency means of egress lighting and internally illuminated exit signs with backup power supplies were provided. However, their efficacy could not be determined through analysis of the actions of the survivors. The accounts of survivors shed no new light on the discussion of emergency lighting and exit sign requirements in the model codes. The initial responses of the individuals most directly affected by this fire seemed to be more directly oriented toward escaping from the room and extinguishing their burning clothing than evacuating the building *per se*. As such, their actions were probably not focused on looking for exit signs or other obvious indicators of the way to exits.

Many of the fourth floor survivors indicated that the corridor was dark, probably due to the heavy smoke produced by the burning aviation fuel not a failure of the emergency lighting. Emergency lighting is required in corridors and means of egress to assist occupants in identifying the path of travel and avoiding egress hazards. However, the lighting levels specified in the model codes are relatively low and generally incapable of overcoming the obscuration of dense smoke.

The merits of different methods of illuminating exit signs and their location have been subjects of considerable debate in the code development community through the years. Even internally illuminated exit signs emit less light than emergency lighting units and are generally installed at or near ceiling level which means that they can be easily obscured by rising smoke.

LESSONS LEARNED

1. **Command and control of a complex interagency emergency response is enhanced significantly by good equipment and well-planned communications procedures.**

   Evansville's new communication system was put to an extreme test and passed with flying colors. This was due in large part to good planning. The telephone system handled a 500 percent increase in call volume during the initial hours of the incident, but all calls got answered. Fireground personnel from the city fire department, township fire departments, local EMS, law enforcement, emergency management, coroner, and public works agencies all were able to communicate effectively if not always efficiently. Those who were not on the new trunked radio system sent representatives to the command post to relay their messages. Despite the magnitude of the incident and the large number of agencies involved in the response, remarkably few complaints emerged in the formal critique. Those agencies which were not part of the new radio system expressed an interest in either acquiring the hardware or developing a means for sharing equipment at disasters.

2. **The use of the ICS complemented effective communication and facilitated a smooth transition to recovery and investigation.**

   The use of the ICS facilitated an effective interagency response to this disaster. Fire service, law enforcement, EMS, emergency management, and public works departments all worked better because of the coordination and communication facilitated by use of the ICS. Moreover, the transition from response to recovery went smoothly because no single agency "owned" the ICS.

Some of the problems which did arise during the incident were linked to self-responding units and individuals not checking in with the Command Post and integrating their responses into the ICS. Although this led to some confusion, and may have contributed to untoward actions by some responders, the Incident Commander was able to overcome this disorganization and produce an effective response.

3. **Rescue operations should be restricted to personnel trained and equipped to deal with the hazards present.**

    The most serious injuries to emergency responders were suffered by personnel who attempted to perform duties for which they were not trained and not properly equipped. Every emergency responder has a role to play at a major incident. Law enforcement personnel can focus their attention on controlling access to the scene to prevent additional people from becoming involved. EMS personnel can triage, treat, and transport the injured. Controlling the fire and locating and removing the injured from danger should be left the primary responsibility of firefighters.

4. **Adherence to minimum construction standards contributes to successful outcomes.**

    No building is designed to withstand the effects of an airplane crash. However, this fire demonstrates that compliance with model building code requirements and adherence to good design practices contributes to successful outcomes. Although part of the restaurant collapsed when the tail section of the airplane landed on it, the hotel sustained relatively minor structural damage. Only those north-facing guest rooms whose windows shattered on the initial impact sustained heavy fire damage. The interior fire was confined to those rooms except where doors to the corridors were left open. Most of the fire damage was caused by the burning aviation fuel, as opposed to interior finishes and furnishings.

5. **The actions and accounts of survivors reinforce the understanding that human behavior in fire emergencies is generally rational, adaptive, and goal-oriented.**

    Despite the sudden and overwhelming danger which confronted the occupants of the Drury Inn, all of the survivors appeared to react in an adaptive and rational way, and evidence suggests that the victims responded similarly.

APPENDIX A

Vicinity Plan and Building Diagrams

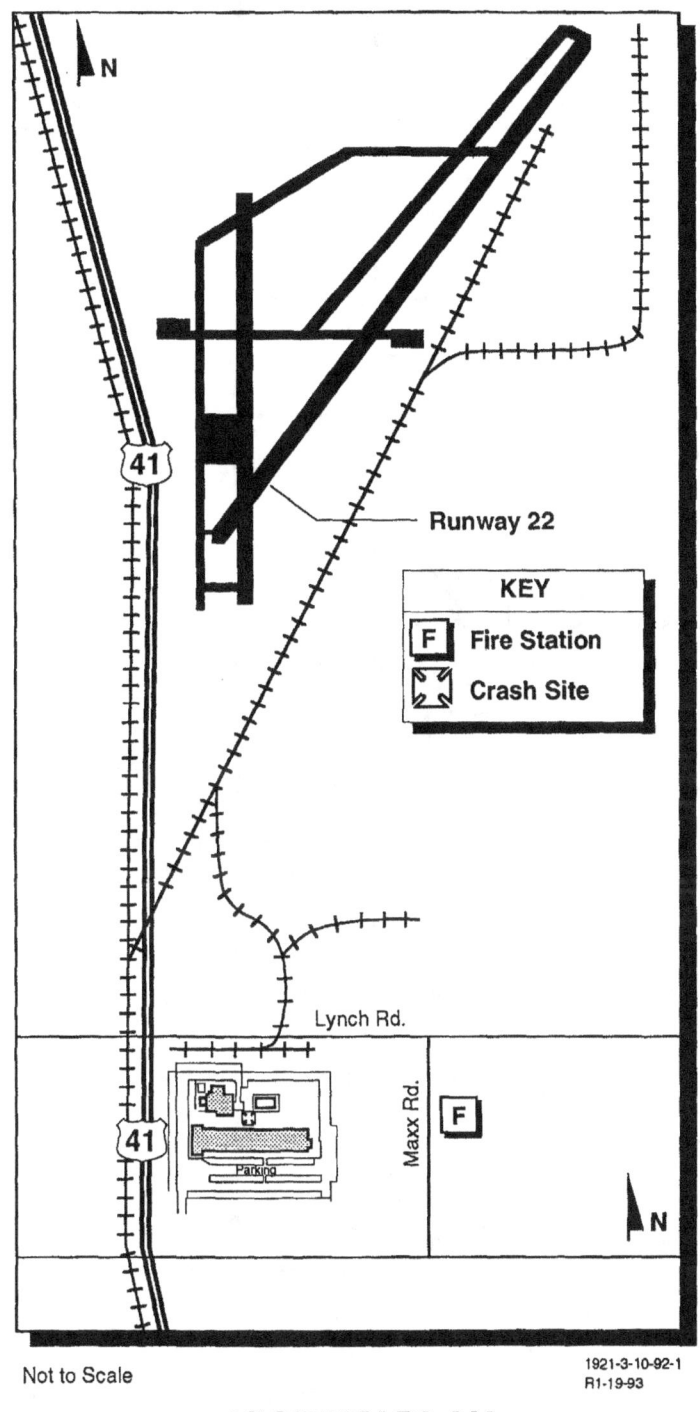

Runway 22

KEY

F — Fire Station

Crash Site

Lynch Rd.

Maxx Rd.

F

41

Not to Scale

1921-3-10-92-1
R1-19-93

VICINITY PLAN

Appendix A (continued)

4th floor

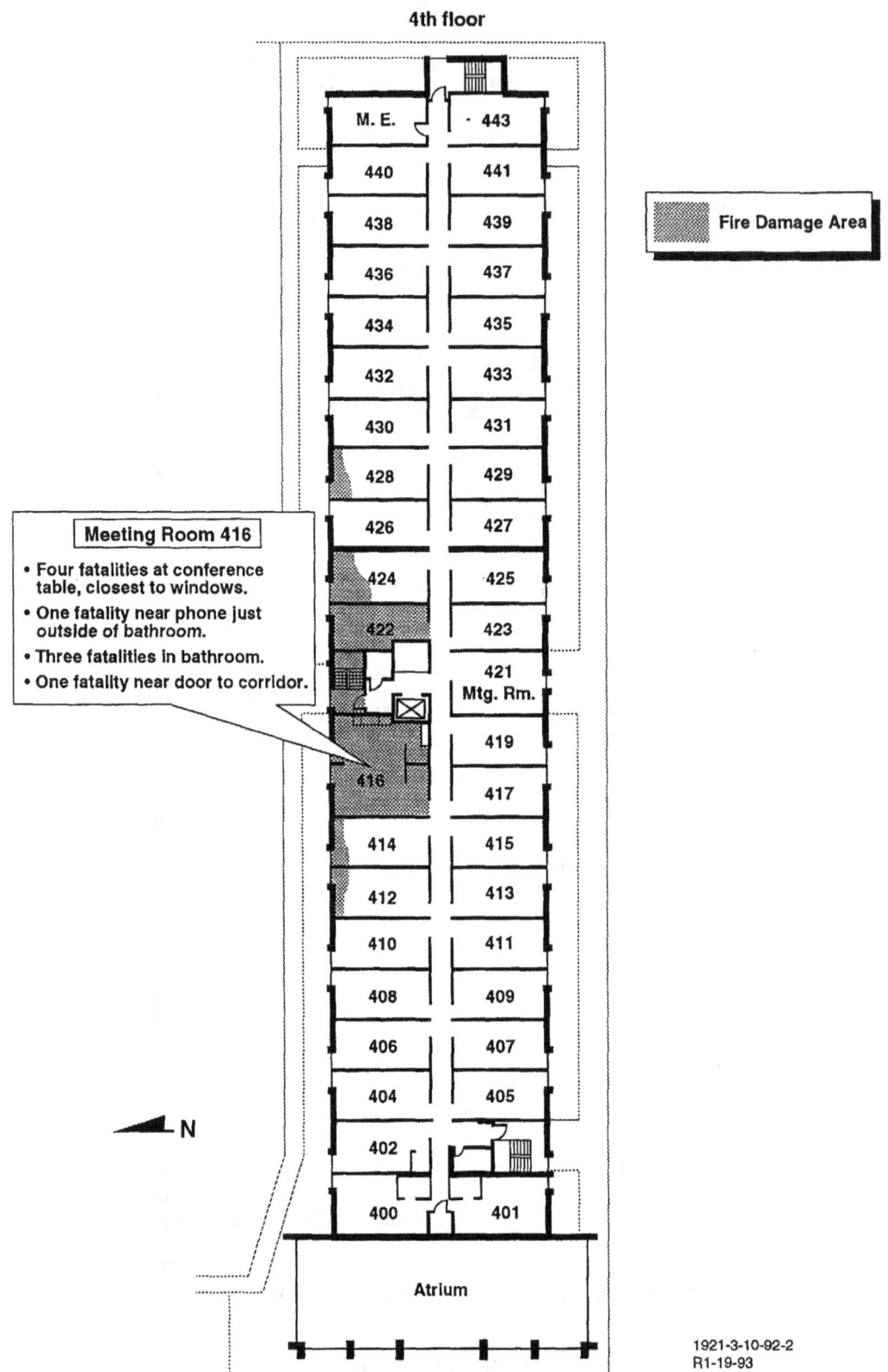

Fire Damage Area

Meeting Room 416

- Four fatalities at conference table, closest to windows.
- One fatality near phone just outside of bathroom.
- Three fatalities in bathroom.
- One fatality near door to corridor.

N

Atrium

1921-3-10-92-2
R1-19-93

Appendix A (continued)

1st floor

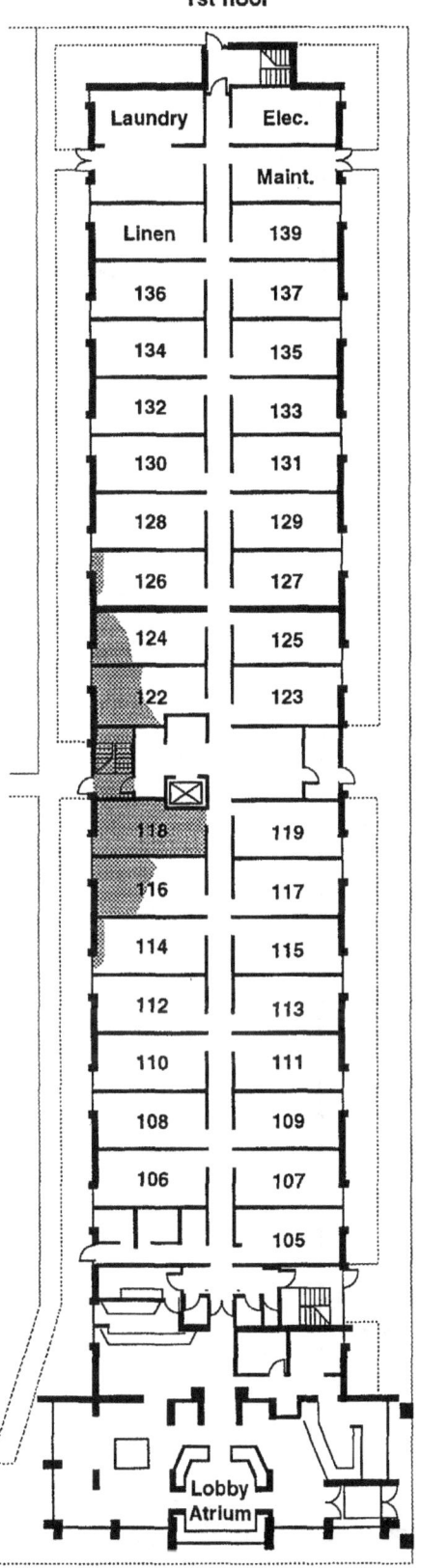

Fire Damage Area

1921-3-10-92-2
R1-19-93

Appendix A (continued)

2nd floor

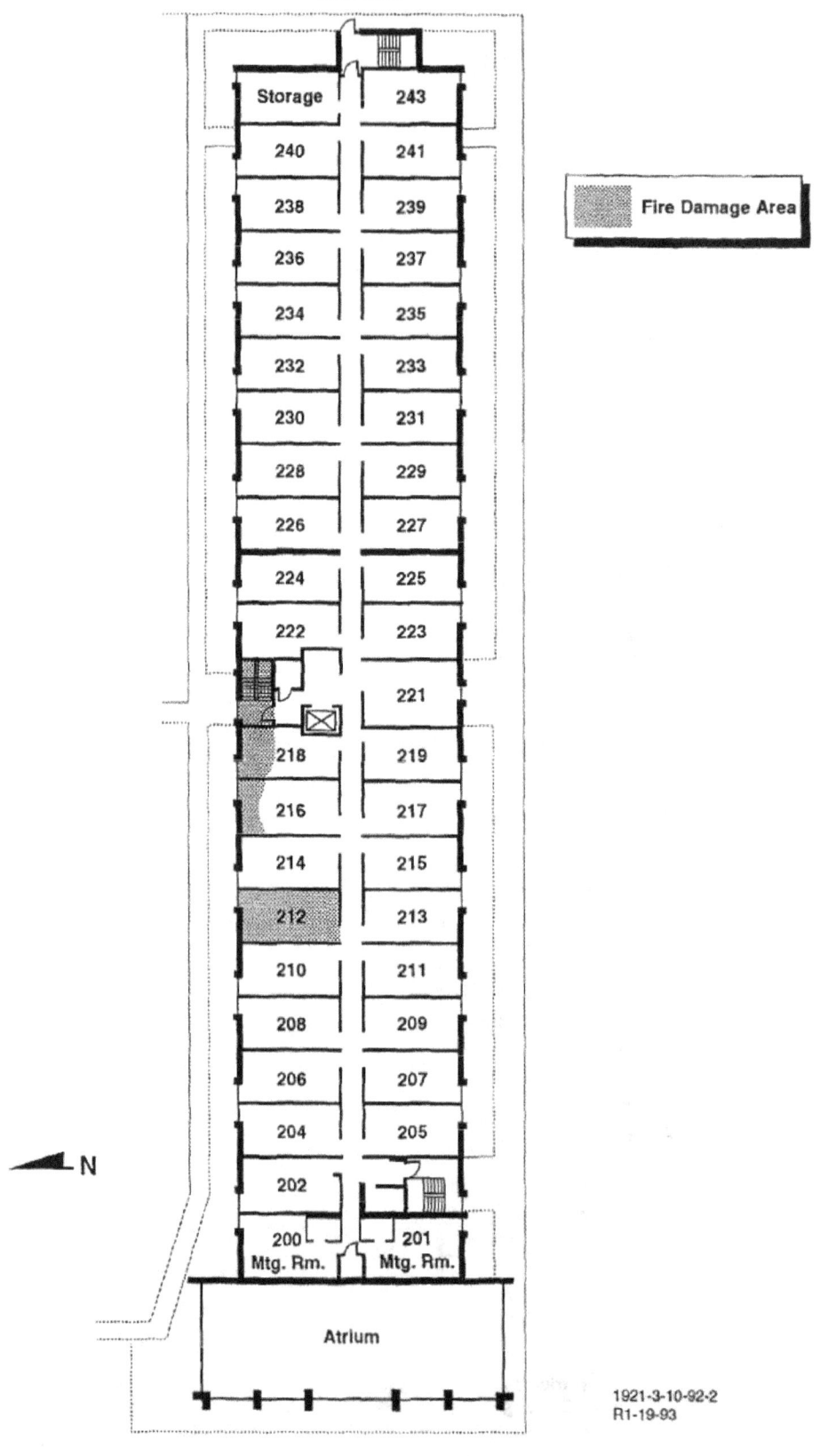

Appendix A (continued)

3rd floor

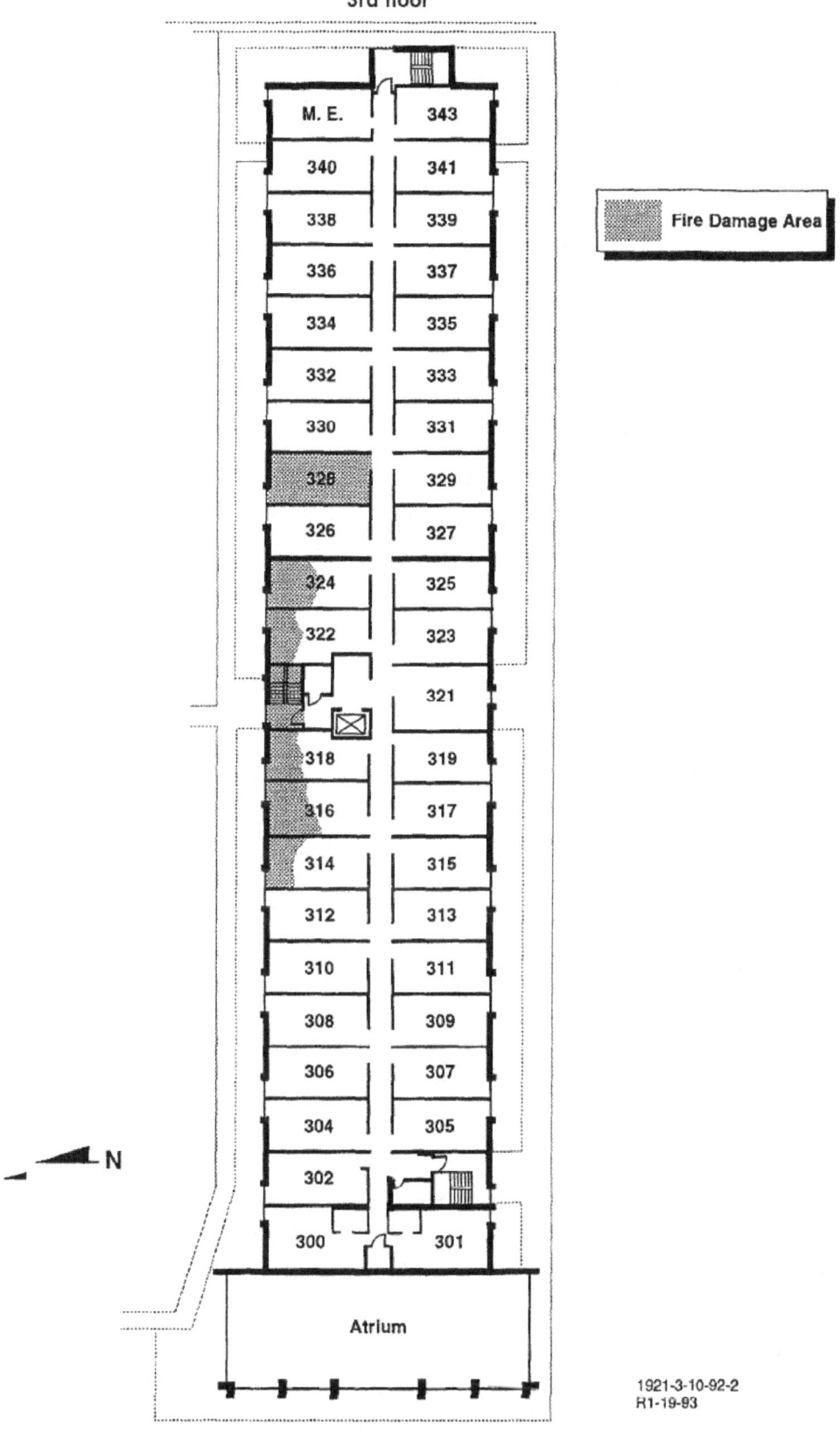

APPENDIX B

List of Responding Agencies

(Provided by the Emergency Management Agency, Evansville-Vanderburgh County, Indiana.)

RESPONDING AGENCIES--FEBRUARY 6, 1992

Evansville/Vanderburgh County Emergency Management Agency: Director--Sherman G. Greer, Assistant--Jane Snelling, Local Emergency Planning Committee--Janice F. Shah

Knox County Emergency Management Agency: Director--Steve Dillon

State Emergency Management Agency: Regional Coordinator--Jack Scott; District Coordinator Emergency Medical Services--Becky Blagrave; State Division Director For Emergency Medical Services Training--Tony Pagano, Chief Disaster Response And Recovery--Phil Brown

Public Information Officer: Christine Terry, Director, EPA

Fire Departments: Evansville, Knight Township, Perry Township, German Township, McCutchanville Fire And Rescue, Gibson Co. Rescue, Airport Safety Officers, Vincennes Township Rescue Squad, Marrs Township--Posey County

Emergency Medical Services: Alexander Ambulance, Bassemeirs Ambulance, Comaier Ambulance, Madison Ambulance, Henderson Ambulance, Warrick Co., McClure Ambulance/Bicknel, Gibson Co. Ambulance, Halter Smith--Vincennes, St. Anthony's Ambulance--Vincennes, One Unit From Illinois

Security/Traffic Control: Evansville Police Department, Indiana State Police, Vanderburgh County Sheriff's Department, Auxiliary Police Officers--Emergency Management Agency

American Red Cross--Evansville

American Red Cross--Knox County

Communications--Telephone: Indiana Bell, Contel Cellular, United States Cellular Mobile Telephone Network, MCI Telecommunications Network

Communications--Radio: Races, React, Evansville Citizens Radio League, Hal & Carol Wilson (Races) Team--Manned The Emergency Operations Center In The Emergency Management Agency Office, Keith Phillips At C.K. Newsome Center

Hospitals: Deaconess, Welborn, St. Mary's

Military Services: Kentucky Air National Guard, Kentucky National Guard, Evansville National Guard, Indiana National Guard, Scott Air Force Base--Bellville, Illinois; Grissom Air Force Base, Air Force Claims Officers (7 From Grissom And Washington, DC), Air Force Hazmat Team, Safety Investigation Board, Adjutant General--Kentucky Air National Guard

Building Commissioner/EMA Advisory Chairman: Roger Lehman

City Engineer: Herb Butler

Southwest Mental Health

County Prosecutor: Stan Levco

Coroner's Office

Central Dispatch: Harold Crooks And Dispatchers

Council Of Churches

Clergy Of Evansville/Vanderburgh County

C.K. Newsome Center: Harvey Taylor

Parks & Recreation: Jim Hadden

Mayor F. Mc Donald & Staff: Cindy Baumgartner Developed Volunteer List For EMA Telephone Bank Work

County Commissioners

City Garage: Bob Brown

County Engineer: Gary Kercher

Ohio Valley Search & Rescue Dog Association: Volunteer Staffing in EMA Office--Dan Casper, Dave Johann, Harold Crooks, Marti Vanada, Janice McNeely, Tina Notter

News Media: Local, State, National, International (Mexico, London) 35 Media Represented

Health Department: Brett Townsend, Volunteer for EMA Office--Sherry Sampson

Convention/Visitors Bureau: Pete Helfrich--For use of his 800 number telephone line and personally manning it for many hours

Emergency Medical Technicians: Volunteers for the Emergency Operations Center and the C.K. Newsome Center--John Williams, Tom Dant, Jean Trulock

National Organization of Victims Assistance (Nova)

Hotels and Their Staff: Ramada Inn, Radisson, Holiday Inn

Alcoa/Warrick Operations: Generators And Lights

Rescoe Rents: Light Towers

Lamasco Equipment

Civic Center Complex: Telephone/Switchboard Operations, Security Officers, Dorothy Burgolt--Night Security

Sigeco

City Water Department

TSF: Portable Johns

Food Donations: Chilli Cooking Teams (For Red Cross): New Hope Baptist Church, Cleaves CME Church, Nazarene Baptist Church, First Ebenezer Baptist Church, Ladies Auxiliary--VFW Posey County, Domino's Pizza, Little Caesar's Pizza, Grandy's, Rax, Kentucky Fried Chicken, Hardee's, McDonald's, Taco Bell, Merry-Go-Round Restaurant, Pantry Convenience Store (Kentucky & Riverside), Denny's

Note: This list is based on the best information available at this time--February 12, 1992. A special thanks to all volunteers--both named and unnamed--who made all the assistance possible to so many agencies and individuals.

J.E.S.

APPENDIX C

Critical Incident Stress Briefing Materials

(Provided by the Southwest Indiana Critical Incident Stress Management Team and the U.S. Air Force Critical Incident Stress Team from Scott AFB Illinois.)

Dear Emergency Services Professional,

The event(s) you recently experienced were more intense than usual, even for emergency workers. These experiences can manifest themselves in a multitude of signs and symptoms:

PHYSICAL REACTIONS

- Fatigue/exhaustion
- Change in appetite
- Difficulty sleeping
- Headaches
- Digestive problems
- Startle reactions

COGNITIVE REACTIONS

- Difficulty concentrating/remembering things
- Flashbacks/intrusive memories
- Feelings of isolation
- Difficulty solving problems or making decisions
- Inability to feel that anything is important other than this incident

EMOTIONAL REACTIONS

- Anger/irritability
- Feeling of helplessness
- Over sensitivity
- Fear
- Loss of memory of event
- Depression
- Guilt
- Emotional numbing
- Anxiety

Although no one can prevent you from having these reactions, the following is a list of suggestions which should help you minimize the impact of them:

STEPS YOU CAN TAKE

- Alternate periods of strenuous physical exercise and periods of relaxation during the first few days. This will help release some of the pent-up physical tensions and reactions.
- Structure your time in order to "keep busy."
- Do not label yourself "crazy," You are having **NORMAL** reactions!
- Talk to people – Talk is the healthiest medicine.

Appendix C (continued)

- Beware of overusing drugs or alcohol to numb the pain. You don't need to complicate the problem with substance abuse.

- Other "mood altering" substances such as caffeine and sweets should be used in moderation.

- **REACH OUT!** Spend time with others. People DO care.

- Keep your life as normal as possible. Do not make any "big" life changes. Do make as many daily decisions as possible. This will help you regain the feeling of control over your life.

- Help your co-workers by sharing your thoughts and feelings and be willing to listen to theirs. Check out how each of you are doing. You will find a lot of "common ground" concerning your thoughts and reactions.

- Give your self permission to feel rotten and share your feelings with others.

- The above symptoms should diminish over time (3 to 6 weeks). If they do not, or if they intensify, you may need to seek additional professional help in dealing with this event. Remember, **YOU ARE HAVING NORMAL REACTIONS TO ABNORMAL EVENTS!**

Our only purpose is to help you deal with your reactions to the stressful event you just experienced. **WE CARE ABOUT YOU!**

SOUTHWEST INDIANA CRITICAL INCIDENT STRESS MANAGEMENT TEAM

OUR TEAM KEEPS YOUR TEAM ON THE JOB!

Appendix C (continued)

CRITICAL INCIDENT STRESS OVERVIEW

- You have experienced a traumatic event or a critical incident (any incident that causes you to experience strong emotional reactions which have the potential to interfere with your ability to function either at the scene or later). Even though the event may be over, you may now be experiencing or experience later, some strong emotional reactions. It is very common, in fact normal, for people to experience emotional aftershocks when they have passed through a terrible event.

- Sometimes the emotional aftershocks (or stress reactions) appear immediately after the traumatic event. In some cases, weeks or months may pass before the stress reactions appear.

- *The signs and symptoms of a stress reaction may last a few days, a few weeks, or a few months, and occasionally longer, depending on the severity of the traumatic event. With understanding and the support of loved ones, the stress reactions usually pass more quickly. Occasionally, the traumatic event is so painful, that professional assistance from a counselor may be necessary. This does not imply craziness or weakness. It simply indicates that the particular event was just too powerful. Think of it as a circuit breaker in an electrical system that disconnects, rather than risking overload to the system. The circuit breaker has done its job by temporarily disconnecting. The system is still intact and all that is needed is for the circuit breaker to be "reset." That can be accomplished through peer counseling, defusing, debriefing and occasionally with the help of a behavioral science specialist.

- Here are some common signs and signals of a stress reaction:

PHYSICAL	COGNITIVE	EMOTIONAL
-fatigue	-blaming someone	-anxiety
-nausea	-confusion	-guilt
-muscle tremors	-poor attention	-grief
-twitches	-poor decisions	-denial
-chest pain	-heightened or lowered alertness	-severe pain (rare)
-difficulty breathing	-poor concentration	-emotional shock
-elevated blood pressure	-memory problems	-fear
-rapid heart rate	-hyper vigilance	-uncertainty
-thirst	-difficulty identifying familiar objects or people	-loss of emotional control
-headaches	-increased or decreased awareness of surroundings	-depression
-visual difficulties	-poor problem solving	-inappropriate emotional response
-vomiting	-poor abstract thinking	-apprehension
-grinding teeth	-loss of time, place or person orientation	-feeling overwhelmed

Appendix C (continued)

-weakness	-disturbed thinking	-intense anger
-dizziness	-nightmares – intrusive images	-irritability
-profuse sweating	-agitation	-fainting
-chills	-shock symptoms	

BEHAVIORAL

-change in activity	-change in speech patterns	-hyper-alert to environment
-withdrawal	-emotional outbursts	-startle reflex intensified
-suspiciousness	-loss or increase of appetite	-non-specific bodily complaints
-alcohol consumption	-change in usual communications	-increase or decrease in sexual functioning
-inability to rest	-erratic movements	-antisocial acts
-pacing		

The Normal Recovery of Normal People Experiencing Normal Reactions to Abnormal Events!

Sane People Involved in an Insane Experience!

THINGS TO TRY

- Within the first 24-48 hours, periods of strenuous physical exercise, alternated with relaxation will alleviate some of the physical reactions.

- Structure your time – keep busy.

- You're NORMAL and having NORMAL reactions – don't label yourself as crazy!

- Talk to people – talk is the healthiest medicine.

- Be aware of numbing the pain with the overuse of drugs or alcohol, you don't need to complicate this with a substance abuse problem – Reach out – people do care.

- Maintain as normal a schedule as possible.

- Spend time with others.

- Help your co-workers as much as possible by sharing feelings and checking out how they're doing.

- Give yourself permission to feel rotten and share your feelings with others.

- Keep a journal, write your way through those sleepless hours.

- Do things that feel good to you!

Appendix C (continued)

- Realize those around you are also under stress.

- Do make as many daily decisions as possible which will give you a feeling of control over your life (i.e., if someone asks if you want to eat, answer them even if you're not sure.)

- Get plenty of rest.

- Reoccurring thoughts, dreams or flashbacks are NORMAL – don't try to fight them – they'll decrease over time and become less painful.

- Eat well balanced and regular meals. (Even if you don't fee like eating.)

FOR FAMILY AND FRIENDS

- Listen carefully.

- Spend time with the traumatized person.

- Offer your assistance and listening ear, even if they don't ask for help.

- Reassure them that they are safe.

- Help them with everyday tasks, like cleaning, cooking, caring for the family, minding the children, etc.

- Give them some private time.

- Don't take their anger or other feelings personally.

- Don't tell them that they are "lucky it wasn't worse" – traumatized people are not consoled by those statements. (It will make them feel guilty for feeling what they feel.) Instead, tell them that you are sorry such an event has occurred and that you want to understand and assist them.

- Don't feel like you have to have all the answers or know exactly what to do. Just being there shows them that you care. (It is okay to ask them what they would like for you to do.)

(Compliments of the Combat Stress Treatment Unit) Major Meyer

Major Duane Meyer, Capt. Jack Little
Critical Incident Stress Team
USAF Med Cetr/SCHA
Scott AFB Illinois 62225-5300
AV 576-7386/87
Commercial 618-256-7386/87

APPENDIX D

Evansville-Vanderburgh County Emergency Management Agency

Evansville Plane Crash After Action Report

EMERGENCY MANAGEMENT AGENCY
EVANSVILLE, VANDERBURGH COUNTY

Sherman G. Greer. Director
Room 18, City-County Complex
1 NW Martin Luther King, Jr. Boulevard
Evansville, Indiana 47708
(812) 426-5602

Jane Snelling,
Administrative Assistant

EVANSVILLE PLANE CRASH - AFTER ACTION REPORT

Date: February 19, 1992

Location: Evansville Vanderburgh County E.O.C.

Critique/Lessons Learned: Sherman G. Greer

1.   Communications

   *   Communications staff accompanied him at all times

   *   Press briefing room set up

   *   Rumor Control established - rumors of 28 dead

   *   Fire suppression - fantastic job done by Airport
       and/or Evansville Fire Department

   *   Law Enforcement - perimeter control established;
       U.S. 41 controlled

   *   EMS response - (looked like ambulance convention)

   *   Support Agencies -

        Red Cross
        Salvation Army

   *   Incident command system established

Appendix D (continued)

INCIDENT COMMAND SYSTEM

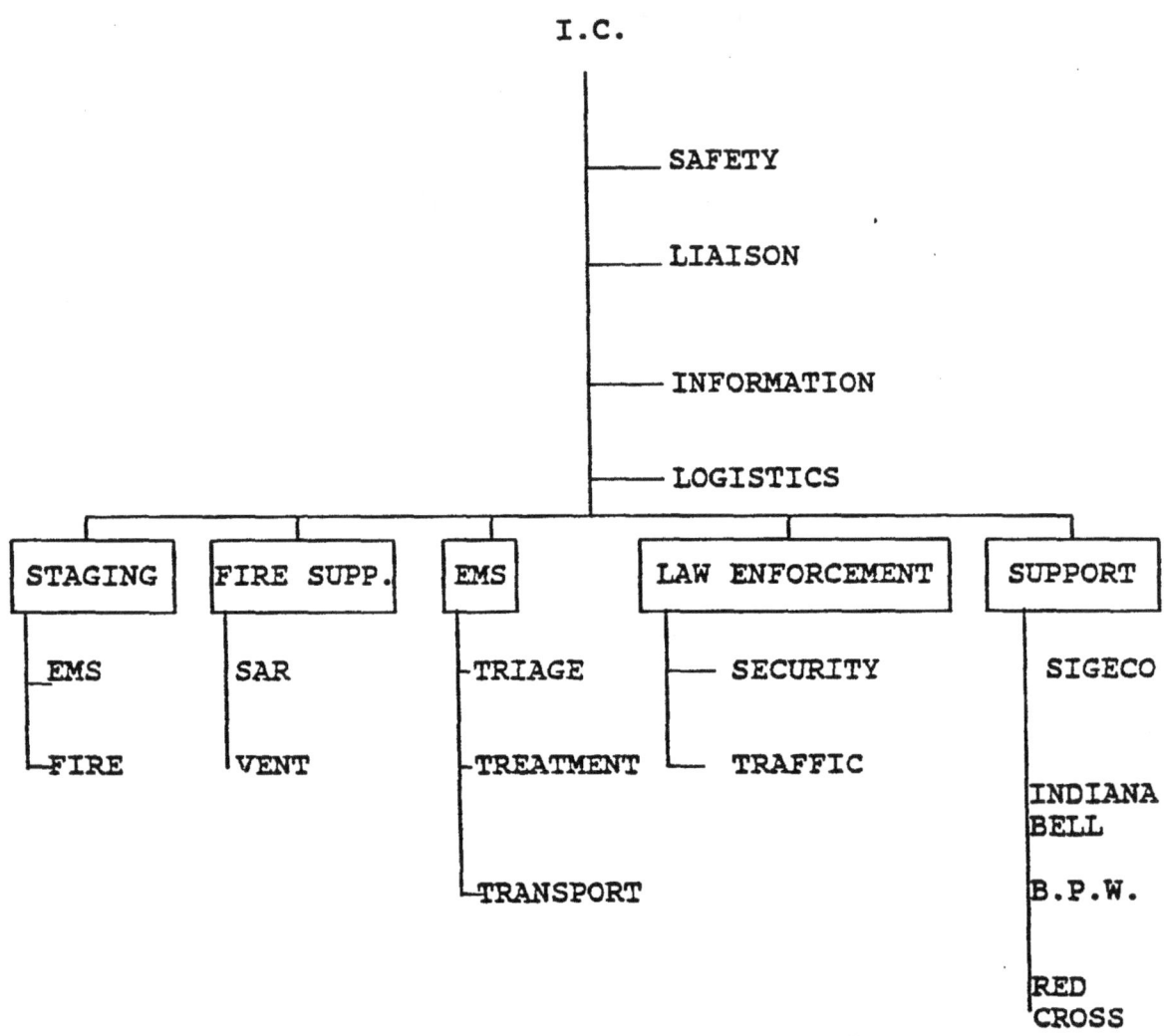

Appendix D (continued)

Introductions: Jack Scott, Area Coordinator SEMA
                  Becky Blagrave, EMS Coordinator SEMA

Red Cross: Trish DeVoy - Everyone worked together real well.
    Rob Goble helped with communications at Red Cross. Red
    Cross lost power and phone service.

Salvation Army: Went to staging area and no one was there.
    (Sherman suggested the Fire Inspector act as staging
    officer). Jesse Storey said we would have to talk.

P.I.O.: Christine Terry - Communications is still important.
    Came to E.O.C. from Fire Dept. Very busy - need for
    portable radios. Too many people at site. Site control
    good.

Deaconess Hospital: Communication from scene to hospital.
    Need designated Family Center. Victims' families had to
    go to all three hospitals. Occupancy full. Need
    community transportation services.

St.Mary's Hospital: Number of victims and need communication
    from scene. Information came in slow. Communications
    between hospital also not good. Best information came
    from Channel 14 live (television).

ALCOA: Unified command structure from all agencies.
    Representatives from all agencies to be at I.C.

Welborn Hospital: Same as other hospital. Utilization of
    resources. Helicopter not called to scene. High level
    of care at scene. Need automatic response to disaster
    scene. Need to update medical plan. CNN met helicopter
    at Lousiville and pilot had no idea what site looked
    like.

REACT: Radio Problems. Gave information to traffic at site.
    Radio interference.

Amateur Radio:

Knox County EMA: Assisted Sherman Greer. Good job and well
    done by all concerned.

Fire Chief Doug Wilcox: Concerns from Fire Dept. - only
    concern was safety - keep unauthorized access from fire
    ground. Tighten up own security in own agency. People
    without proper training and equipment become part of the
    problem. If we need assistance we will call. Then you
    will be directed where to go. Don't come to be part of
    the problem. All people need to be accounted for. I.C.
    Systems will be in all incidents in Evansville till

Appendix D (continued)

Sherman has arrived at the scene. Orderly transition was and would be needed at any disaster.

Law Enforcement - Captain Cradock (Sheriff's Dept.): 800 meg system good.

Evansville Police Dept. - Chief Art Gann: Additional ICS training by law enforcement security part of scene. Fire and EMS was allowed access. After scene was stable coroner and military always allowed in. Law enforcement went in with no air tanks. Need to back out and let fire dept. take over. Coroner and 800 meg. worked well. Radio good. Indiana State Police need access to 800 meg system. Whether ISP or EPD buys - it is needed.

Indiana State Police Officer Taylor: Need 800 meg. system for ISP aircraft investigator. No problems. Remove unneeded people from scene. Situation board as needed. (Sherman Greer comment: IC area needs to be identified)

EMS - Max Kleugh (Alexander Ambulance): Thanked everybody that helped at scene. 35 ambulances arrived within 30 minutes. 800 meg. needed for EMS systems. No contact with incident commanders. Transported private individuals within 20-25 minutes. Information to hospital important. Direct radio contact to hospital. Releasing within 30 minutes. T&T went ok. Deaconess Hospital was control hospital. Internal session help at Alexander with good plans for the future. (Sherman, Medical Task Force will get together shortly. Joint command type system will be used in future). Incoming services went to Alexander private channel. Will have additional frequency in the future. Also, once people are released, they need to leave and not hang around. ID system will be adopted.

Central Dispatch: Traffic was too much. Need other numbers for continuous request for additional assistance. Need people to go to disaster frequencies quicker. Airport security had 800 meg system but was not trained. Went to "Disaster 1" frequency right away. Could not have done without 800 meg. system. Need perimeter control on different frequencies. Need information number so people will not call dispatch.

Coroners'Office: News Releases were a problem. One location for all news media. Items should be given to Sherman Greer for media release. Bad information was given out. All information should go through one person. Coordinate all press releases. "28 people killed" - did come from coroners' office. First call at 10:01 AM. Calls from Japan, Britain, and Canada. Everything done

Appendix D (continued)

within 24 hours.

Airport: Briefly commented - radio system being installed.
    Communications to Central Dispatch worked without any
    problems.

Public Works: Could have been worse weather conditions.
    Large trucks were used as barricades on US 41 North and
    South. Other barricades were brought in on request.

County Response: Time will be a problem. Do not respond
    unless called. CNN showed county highway truck on
    television. Traffic control liability.

Comaier Ambulance: Every EMS person should have gone to
    IHERN or common frequency. (Sherman Greer: 155.34
    should be main frequency)

Bassmeier Ambulance: Responded late and not directed where
    to go. Impressed with the way everything went.

Mr. Gamble: We don't pay Sherman enough for amount of stress
    in his job. Gave him credit for his efforts.

State EMS: Good EMS response. State serves as backup to
    local event.

Jack Scott: Communications and ICS, when you get involved in
    exercise everybody should participate. Vanderburgh Co.
    getting good reviews.

Other comments: Sherman Greer needs 800 meg radio.
    Convention Bureau donated 800 number telephone line.
    Cellular telephone companies donated 6 telephones.
    Drury Inn Scholarship Fund $25,000.

In summary, everybody did well and the good hard work will
continue. The concept of the plan is falling together. Keep
working and keep improving.

 Start: 10:00 AM
 Finish: 11:36 AM

 SGG:jfs

APPENDIX E

Photographs

Photo by Mark Chubb

View of crash site looking south toward Drury Inn. Plywood covers many of the windows broken by the blast from the crash.

Appendix E (continued)

Photo by Mark Chubb

This photograph taken from the roof of the Drury Inn shows the crash site and damage to Jojo's Restaurant. Note the aircraft's main landing gear near the center of the picture.

Appendix E (continued)

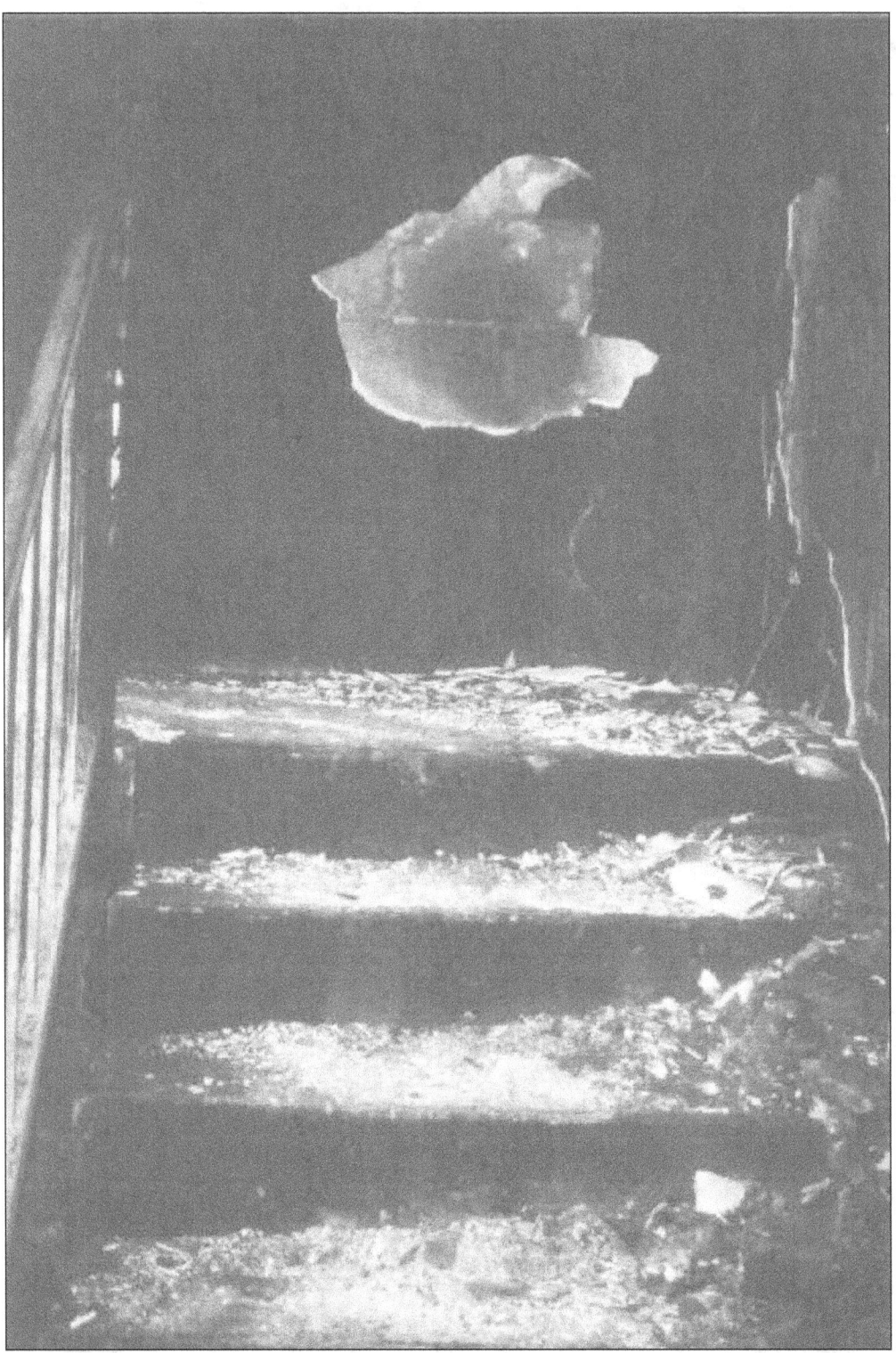

Photo by Mark Chubb

The center stairway (on the north side of the Drury Inn) was heavily damaged by the fire when the windows broke. Note the heavy debris on the stairs and the wall from concrete spalling.

Appendix E (continued)

Photo by Mark Chubb

Even on the most heaving damaged four floor of the Drury Inn, fire damage from flame rollover was limited to locations where corridor doors remained open during the fire. The damage in the foreground is outside room 416.

Appendix E (continued)

Photo by Mark Chubb

In room 328, heavy fire damage resulted when the windows shattered and
fuel was sprayed throughout the room. However, the concrete masonry
construction between guest rooms prevented the fire from spreading to
adjacent spaces.

Appendix E (continued)

Photo by Mark Chubb

Automatic closing fire doors at the west end of the corridors prevented smoke from spreading into the atrium. However, one of the occupants of room 416 became trapped on the balcony after the doors closed behind him when the fire alarm system activated.

Appendix E (continued)

Photo by Mark Chubb

Although the center stairway (on the north side of the Drury Inn) was heavily damaged by the fire, the fire door prevented it from spreading into the corridor.

Appendix E (continued)

Photo by Mark Chubb

In rooms where windows remained unbroken, no fire or smoke damage was evident.

Appendix E (continued)

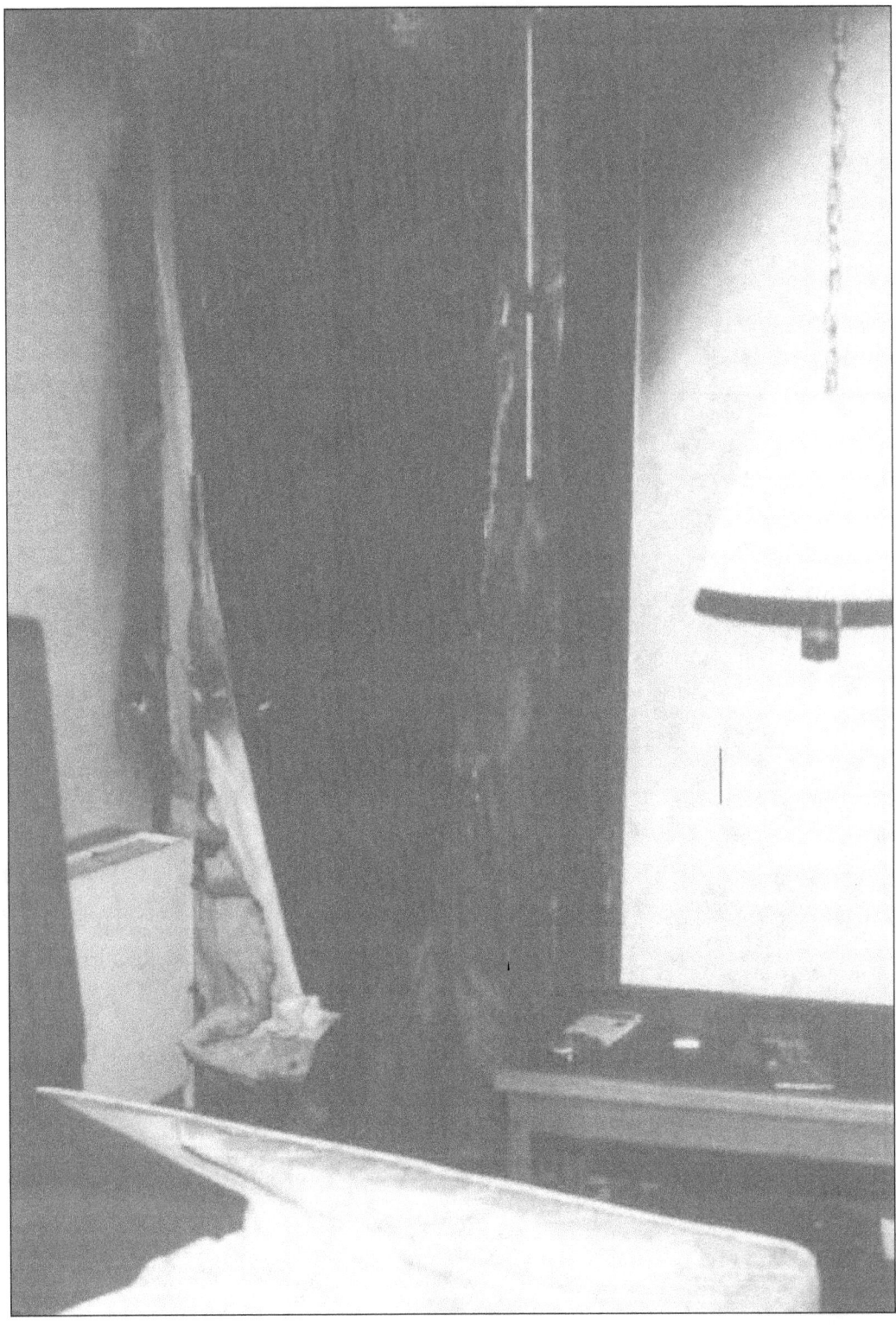

Photo by Mark Chubb

Even where windows were broken, fire damage was minimal in cases where only a limited amount of aviation fuel sprayed in.

Appendix E (continued)

Photo by Mark Chubb

The first floor corridor (looking east) remained free of smoke and flame. Fire resistance rated room doors held the fire in check.

www.ingramcontent.com/pod-product-compliance
Lightning Source LLC
Chambersburg PA
CBHW081625170526
45166CB00009B/3101